Beginner / Intermediate Guide to Bladesmithing

Bladesmithing Compendium for Beginners and Intermediate Level Knife Makers

Wes Sander

Table of Contents

BOOK 1: BLADESMITHING 6

INTRODUCTION 10

CHAPTER 1 GETTING STARTED AS A BLADESMITH 13

    SKILLS ESSENTIAL FOR A BLADESMITH 16

CHAPTER 2 MAKING YOUR FIRST KNIFE THROUGH STOCK REMOVAL 18

CHAPTER 3 MAKING A KNIFE THROUGH SMITHING 21

CIIAPTER 4 ADDING A HANDLE, STRENGTH AND FLAIR TO YOUR KNIVES 24

    HEAT TREATING YOUR KNIFE 24

    TEMPERING YOUR BLADE 25

    ETCHING YOUR KNIFE 27

[BONUS 1/3] MAKING YOUR OWN TOOLS 29

    MAKING YOUR OWN FORGE 29

    MAKING YOUR OWN PAIR OF TONGS 30

    MAKING YOUR OWN HAMMER 32

[BONUS 2/3] A DISCUSSION ABOUT DAMASCUS 34

[BONUS 3/3] A PEEK INTO SWORD SMITHING 38

HOW IS FORGING A KNIFE DIFFERENT FROM FORGING A SWORD? 40

CONCLUSION 42

BOOK 2: INTERMEDIATE GUIDE TO BLADESMITHING 43

INTRODUCTION 46

CHAPTER 1: COMPREHENSIVE GUIDE TO KNIFE MAKING 48

CHAPTER 2: COMPREHENSIVE GUIDE TO SWORD-MAKING 66

CHAPTER 3: BLADE-MAKING STEEL 85

CHAPTER 4: COMPREHENSIVE GUIDE TO FORGING DAMASCUS 94

CHAPTER 5: LOST WAX CASTING 103

CHAPTER 6: JAPANESE BLADE 106

CHAPTER 7: GUIDE TO BUILDING A SIMPLE FORGE 112

CHAPTER 8: ARRANGING THE SHOP 114

LEAVE A REVIEW?..116
CONCLUSION: SELF-EDUCATION........................117

Free Bonus for the Readers

Wouldn't it be amazing if you could sell the knives that you make for a profit? Selling your knives can even support your bladesmithing hobby.

However, people often don't know where and how to sell their knives. To solve this problem I have a free bonus for you.

Get the 'Bladesmith's Guide to Selling Knives' e-book for **FREE!** It is a guide on how to sell knives online and in bladesmithing expos, without getting cheated.

All you have to do is go to this link, **http://eepurl.com/gbq6Sb**, and enter your name and e-mail ID.

By signing for the book, you are also signing up for my bladesmithing e-mail list. I occasionally share useful bladesmithing tips and content with this e-mail list.

BLADESMITHING

A DEFINITIVE GUIDE TOWARDS BLADESMITHING MASTERY

Copyright © 2018 by Wes Sander

All Rights Reserved

DISCLAIMER

This document is geared towards providing exact and reliable information in regards to the topic and issue covered. The publication is sold with the idea that the publisher is not required to render accounting, officially permitted, or otherwise, qualified services. If advice is necessary, legal or professional, a practiced individual in the profession should be ordered.

This is in terms of a Declaration of Principles which was accepted and approved equally by a Committee of the American Bar Association and a Committee of Publishers and Associations.

In no way is it legal to reproduce, duplicate, or transmit any part of this document in either electronic means or in printed format. Recording of this publication is strictly prohibited and any storage of this document is not allowed unless with written permission from the publisher. All rights reserved.

The information provided herein is stated to be truthful and consistent, in that any liability, in terms of inattention or otherwise, by any usage or abuse of any policies, processes, or directions contained within is the solitary and utter responsibility of the recipient reader. Under no circumstances will any legal responsibility or blame be held against the publisher for any reparation, damages, or monetary loss due to the information herein, either directly or indirectly.

Respective authors own all copyrights not held by the publisher.

The information herein is offered for informational purposes solely, and is universal as so. The presentation of the information is without contract or any type of guarantee assurance.

The trademarks that are used are without any consent, and the publication of the trademark is without permission or backing by the trademark owner. All trademarks and brands within this book are for clarifying purposes only and are owned by the owners themselves, not affiliated with this document.

INTRODUCTION

The art of bladesmithing has a legacy which outlasts any civilization, with its roots in the Stone Age, where blades were made out of stone and flint, by banging them against each other. This tool was pivotal in the success of humans against other species during the Stone Age. With the advent of the Iron Age came the first metal knives and swords. Bladesmithing took off and the armies with the sharpest swords ruled the lands. Around the time of the invasion of Syria, Damascus was found, sharp and hard to such a degree that it cut through the Spanish invaders' swords! Thus throughout history there have existed blades and bladesmiths, and you are about to be a part of that legacy.

If you are reading this then you are probably interested in making knives, and you are ready to invest time and money into it. *'Bladesmithing: A Definitive Guide Towards Bladesmithing Mastery'* has been designed to fulfill your desire to learn bladesmithing.

This book has been made to avoid as much of the complicated jargon related to bladesmithing as possible, and the whole idea is to keep the book simple.

Something that I recommend is to make products that you can actually sell. So you can actually **MAKE MONEY** while following this book!

Here are the answers to some questions you might have about this book:

What is this book about?: This book is a guide on how to start bladesmithing with the barest of tools.

Who should read this book?: If you are interested in making knives and swords of any kind, you would want to read this book. Anyone looking for a hobby should also buy this book!

Do I need to have tools before I read this book?: Only the barest minimum are required. The rest you can make or acquire along your journey.

What am I going to learn in this book?: I teach how to start bladesmithing from scratch and how to make knives and even add detail to them.

The first chapter deals with the basic progression into forging your first knife. The second and third chapters discuss actually making the knife, while the fourth chapter discusses heat treating and tempering the knife, as well as adding the handle. The goal of this book is to make sure that you spend as less time reading, while applying the knowledge in this book, i.e. I want you to get through this book quickly and start making knives.

Be sure to check out the **BONUS** section where you can learn to make even the tools for yourself, at the end of this book.

Bladesmithing is a hobby that you will cherish for a lifetime and never get tired of. I hope you enjoy and take full advantage of this book.

CHAPTER 1
GETTING STARTED AS A BLADESMITH

So now that you have decided that you want to bladesmith, what's next? Smithing is a process that takes time, and you need to learn throughout your smithing journey. The main focus of this book is to teach how to make your own knife, and this chapter is about how we are going to go about it.

Before we start let us go through the body of a knife, and go through the certain distinctions that you have to make to turn a piece of metal into a knife.

These are:

Blade: This is the cutting part of the knife. But sometimes the whole knife is also referred to as the blade.

Handle: The part of the knife that you hold. It is generally made out of wood.

Spine: The back of the knife.

Choil: A small indent in front of the edge, which marks the end of the edge.

Tang: The part of the blade inside the handle.

Cheek: The flat part of the blade.

Bevel: The slope towards the edge in a knife.

Pay special attention to these terms, as they are essential to the instructions given in the next chapters.

Now that we know the parts of a blade, let us move to the preparation for bladesmithing. The first step would be to get the tools. If you want to make the tools on your own (besides a hammer and an anvil) go to the bonus section at the end.

The type of material you work with is important too. Wood and plastic can be the starting point for most hobbyists as they can be given shape easily, as handles and such. High carbon steel is required for building a knife. Avoid more advanced materials like Damascus and titanium as they are harder to machine.

Set up a vice, anvil, sandpaper, hammer, center punch, file, saw and a drill. The steps given below mention the use of other tools, but they all can be substituted with the ones given above. As mentioned earlier you can also sell the knives that you make and buy more tools as you progress!

You don't need a lot of space to get started either. However be sure that you don't have the risk of a fire hazard wherever you're setting your workshop!

The second step is to ready your mindset. Bladesmithing requires time. You have to develop muscle memory and acquire knowledge of different smithing techniques to get better at it.

The third step is to make. Consider making a pair of tongs, if you don't have them already. It will help you get a hang of the smithing process before you attempt making an actual blade. The guide to making tongs is also available in the bonus section at the end.

After you have handled some smaller projects, it would be a good time to start smithing blades. There are two main options that you can choose, smithing the knife or making the knife through stock removal. In stock removal, i.e. you cut/grind material off the base piece to make your knife. This method is beginner friendly. However you can also choose to smith your knife by heating and banging it with a hammer, and at this point you are probably skilled enough to do either.

Your first knife need not be perfect. But just going through the process of blade making can increase your mastery.

SKILLS ESSENTIAL FOR A BLADESMITH

As you practice the methods given in the next chapters, you will acquire mastery. However, only knowing the process well is not enough for mastery. There are certain other characteristics that are desirable in a bladesmith and they are stated below.

Safety: Understanding the risk and harm that can be caused by the equipment and being careful around the steel. Keep a fire extinguisher and first aid kit handy, and always wear protective gear. Be careful and inspect all tools before using them. Rest if you get too tired. I would recommend going through safety manuals as well. There are several such manuals that are published on the internet for free.

Being one with the steel: You should not fear the heat and the prospect of getting cut. It is natural for a bladesmith to be burned or cut while making knives. You simply cannot be afraid.

Strength: This is especially important if you prefer forging knives. While you don't need a ton of strength to make knives, you need a certain amount of sturdiness to handle the equipment and the blade with stability and dexterity.

Knowledge: For a bladesmith the journey is never over, as is the amount of knowledge that the person can gain. Ideally, to continue becoming better, you need to invest in yourself by reading books and doing courses.

Dexterity and skill: The mark of a master bladesmith is the unique flair that he/she adds to his/her products. This flair can come from the precise way the bladesmith handles the tools, to the sheer brilliance of design that one can bring into the game.

CHAPTER 2
MAKING YOUR FIRST KNIFE
THROUGH STOCK REMOVAL

Stock removal is ideal for your first knife as it is easier, faster and cheaper than forging a knife. This chapter only deals with making the blade. For heat treatment, tempering, and the handle go to Chapter 4.

1. First get an idea of the knife that you want to make, think about the handle, the shape of the blade and other characteristics of the knife. Then draw an outline of the knife on a piece of paper. Make it as well as you can, but it doesn't have to be perfect. Be sure to mark where the handle starts (the choil) in the outline.

2. Next, paste the outline, or trace it onto a piece of steel. I would recommend O1 steel. However, other steels like the 1080 high carbon steel will also work well.

3. Then use an angle grinder or a saw, to grind away the excess metal till you get very close to your markings. Be sure to make every big cut as slow as possible. This will reduce the heat generated and will protect your equipment.

4. After achieving the basic shape of the knife, sand off all of the leftover metal with sandpaper or a belt sander. After the process is complete, you should be left with the metal version of your knife outline. This is where we move on to the step where it becomes a real knife, by sanding the knife to get a bevel and an edge.

5. Trace the bevel outline that you want to have onto the knife. Use a file or belt sander to make the bevel. The key is to make the bevel as smooth as possible, by making light and swift strokes through the file or sander.

6. Set up the knife with a drill press to drill holes (ideally 2 to 3) into the tang, from where you can attach the handle that you will be making in Chapter 4. Be sure to use a coolant of some kind on the blade while drilling.

7. Now we add the edge, using a whetstone, again the strokes have to be light, to ensure an even edge.

8. Heat treat and temper the blade. (Chapter 4)

9. Attach the handle. (Chapter 4)

10. Finish up the blade using a belt grinder/sandpaper.

At this point you should have successfully made your first knife. Congratulations! Why don't you try and sell this piece of work?

SIDENOTE: If you are enjoying this book, can you **leave a review** for it on Amazon? It will really help my book.

CHAPTER 3
MAKING A KNIFE THROUGH SMITHING

This chapter provides steps to make a knife using the forging/smithing technique. For heat treatment, tempering, and attaching the handle go to Chapter 4.

I would recommend you to start out with 1084 steel. While 1095 high carbon steel is traditionally used in forging, 1084 steel can give superior performance for a beginner due it being easier to work with. It is also cheaper.

The major disadvantage of high carbon steel is its vulnerability towards rusting, it is therefore advised that you polish and use methods to keep rust off your steel.

1. Heat the steel. Steel becomes very soft when it is hot so it is essential to heat it. It also becomes easier to work with. You can gauge how hot the steel is by comparing the color of your forge with the color of the metal, i.e. the steel is ready to be shaped once it turns bright orange. It is important to constantly heat the steel, as hot steel flows more easily and the hammer stroke will shape the blade more uniformly.

2. What you should aim at initially is to differentiate between the handle and the blade. You can do so by banging it around the edge of an anvil. Make sure to mark the choil by giving the hot steel a small blow at the edge of the anvil. You might get tired of swinging the hammer so instead of making brutal hits, try and do a more rhythmic hit.

3. Hit the hammer onto the metal such that the blade metal becomes pointier. Usually, the direction of hitting is towards yourself. As the metal is soft, it will start going sideways. Due to this, you want to make sure that you constantly flatten the belly of the blade by hitting it against your anvil after every stroke for making the point.

4. To work the bevel of the blade, hit it with the hammer against the anvil, such that the hammer moves away from you after every stroke. Then we shape the blade to a more sharp looking edge, by hitting it with the hammer with a slight angle.

5. Instead of drilling holes, it is much quicker, easier and cheaper to 'hot punch', i.e. use a center punch and the hole in your anvil to punch in holes into the handle of the knife, while it's still hot.

6. Constantly compare the outline of your knife, and shape it till you are satisfied.

7. Give an edge to your knife with an angle grinder. You can also use a stone such as the JB8 sharpening stone.

8. Heat treat and temper the blade. (Chapter 4)

9. At this step the metal body is ready, and it is time to attach a handle and finish off the blade. (Chapter 4)

10. Finish up the knife with a belt grinder/sandpaper.

CHAPTER 4
ADDING A HANDLE, STRENGTH AND FLAIR TO YOUR KNIVES

Now that you are finished making yourself a brand new blade. It is time to ready it for actual use and add a bit of flair to it.

HEAT TREATING YOUR KNIFE

Heat treating involves heating the knife at a certain temperature and letting it cool slowly. Heat treating allows the molecules in the blade to rearrange and become harder. Heat treating and tempering the blade is the final rite of passage for your blade, before you add the handle and finish it off. I would recommend using an oven at a set temperature for heat treating your knife.

Note: Always use protective and heat resistant gear when heat treating and tempering.

1. Heat the entire blade to a hot orange. To heat the whole blade you might need to shift it back and forth a few times, so keep the tongs handy.

2. After the blade is heated up, quickly quench it in oil (be careful of smoke, use any kind of cooking oil), the blade is quenched specifically in oil because it has the least amount of chance to crack when heat-treated in oil. With water, there is always a chance of cracking, but this is left to your discretion.

3. At this point, you need to test for the hardness of the blade. Run a file over it. If the file marks the steel, then the steel is not hard enough, and there was a problem in the heat treating process that you have to identify and correct. If the file does not mark the blade, then the blade is actually harder than the file, and has now finished hardening.

TEMPERING YOUR BLADE

The tempering is done for the blade to acquire a bit of softness, so that it does not break under pressure. This is done through heating the blade at a steady temperature for around three to four hours. This temperature is different for different kinds of steel. For the recommended steels (O1 and 1084), the temperatures are 350 - 400°F (177 -204°C) for O1 steel and 500°F (260°C) for 1084 steel.

1. Right after quenching, place the blade in an oven with the temperature set according to the steel.

2. After three hours, cool the blade in water.

MAKING THE HANDLE

1. Take two pieces of wood as the handle and mark the tang outline on them. Drill holes on them, corresponding to the holes in the tang of the blade.

2. Use epoxy to seal one piece of the handle with the tang and put in rivets with some epoxy on their tips. Now add epoxy on the other side of the tang and attach the other piece of the handle with the tang and rivets. Use clamps to hold down the handle and tang as firmly as possible, and leave it for 24 hours to cure.

Epoxy is used because it uses a chemical reaction to seal, and is much stronger as compared to a normal adhesive. Be careful with epoxy and make sure everything is right before the epoxy sets. Once it sets it will be very difficult to detach the pieces from the tang.

3. After curing, wrap the blade with masking tape and start sanding the handle using the belt sander. Take your time and make the handle's shape as smooth as you like.

4. After you are finished sanding, it is time to polish the handle using veneer. I would also recommend linseed oil to polish the handle, if you don't have veneer.

ETCHING YOUR KNIFE

This is a guide to DC etching your knife. The other method is acid etching, but it is an advanced technique that is used on knives made out of more complex metals like Damascus. I will be discussing acid etching in my mail list so be sure to go to the Shop and Resources page of this book and sign up for my mailing list!

Note: Before any etching, you have to make sure of your safety, with gloves and glasses.

To do the etching you will need a current source (in our case a 9V battery), two wires, an electrolyte (salt water solution), a piece of cotton and four wire clamps. You will also need a premade design such as a sticker or a piece of paper with the design cut into it. This will allow only the cutout part to be etched.

1. Warm up your etchant (in this case salt water) and knife to your maximize the etching process.

2. Connect the wire clamps to each end of the wires. Connect the clamps of one wire to the positive terminal of the battery and the knife. Connect the clamps of the other wire to the negative terminal of the battery and a piece of cotton dipped in salt water.

3. Gently place the cotton on the sticker/paper. This will start a chemical reaction that will eat away at the metal. Be sure to swab the cotton over the whole etching area, to allow for even etching. You might have to replace the cotton a few times before seeing satisfactory results.

4. Once you are satisfied with the etching, disconnect the wires from the batteries, and remove the sticker/paper from the knife. Clean the etched area.

[BONUS 1/3] MAKING YOUR OWN TOOLS

It is not necessary to invest a whole lot of money into your workshop. This section is a guide on how to make your own tools and furnace. It is important to remember that to make these tools you will still need an anvil and a hammer.

MAKING YOUR OWN FORGE

It is possible to make a small forge out of fire insulating blocks. You don't need to buy a forge to start making knives, however it will be useful to upgrade as you continue to make knives.

For this build we use three fire insulating blocks, a few bricks, a blowtorch, and a drill.

1. Stack two fire insulating blocks on top of each other and draw out a circular opening for the forge at the center face of the stack. Do the same for the back face of the stack.

2. Carve out a cylindrically shaped hole onto the stack, by joining the circles in the front and back, and then cutting through the markings.

3. Drill a hole slightly bigger than the nozzle of your blowtorch, on the side of the stack. Insert the blowtorch through the hole into the cylindrical cavity. Make sure that you do not insert the air inlet along with the nozzle.

4. Put two bricks on top as well as the bottom of the forge, to keep it from moving.

5. The back of the forge can be blocked by the third fire insulating block, but you can remove it if you are heating something longer than the forge itself.

Test out your new forge, by lighting up your blowtorch. The reason a blowtorch is used as compared to a solid fuel like coke or wood is because of the simplicity and control over temperature that a gas-based fuel source provides. This is particularly useful in case you want to do multiple things like smithing, heat treating and tempering.

MAKING YOUR OWN PAIR OF TONGS

Making a pair of tongs is the easiest smithing project that you can do. All you need is a forge, two iron/steel rods and a rivet!

1. Heat up one end of each of the two rods in a forge. Take them out and beat the hot ends with a hammer to slowly shape their cross section as a rectangle.

2. Bend the flat ends into a C shape by beating them around the conical part of your anvil. Bend the ends of the C-curves in the opposite direction to make the gripper part of the tongs.

3. Heat up the rods again and hot punch a hole slightly bigger than the rivet you have.

4. Place the rivet through the hole in one of the bars and beat the top of the rivet to insert it properly. Do the same with the other rod.

5. Now you should have the clenching mechanism ready. Use the tongs a couple of times, to ensure that the rivet is not inserted too tightly, and that the mechanism is smooth.

6. Now that you are satisfied with the mechanism, flatten the ends of the rivet that are projecting out of the rods on either side. Start flattening from the outer end of the rivet and work your way in.

7. Heat up the tongs to a bright orange and then quench them in a bucket.

8. Using an angle grinder/saw, cut off the bars to the length that you find comfortable.

With this you have got yourself a new pair of tongs!

MAKING YOUR OWN HAMMER

To make a hammer, you require a forge, a block of metal, wood, and a hammer. Hence, it only makes sense to make one, if the new one has a different functionality.

1. Take a piece of steel, according to your requirements. Measure the center of the piece on either face and mark it by punching it with a center punch and hammer.

2. Heat the piece to a bright orange in the forge. Using the center punch, punch through the mark. Be sure to remove and cool the center punch constantly. Increase the frequency of cooling as it gets deeper into the hammer head. This is done so that the center punch and the head do not fuse together.

3. Flip the side and continue the same process. Heat up the head if it starts to cool. Do this until the metal in the center falls off.

4. Heat up the hammer head and insert the center punch. Hit the center punch through the head, so that the center gap increases to the desired handle size.

5. After getting a big enough gap, take a piece of wood of suitable length and sand it till its width is equal to the gap. Apply epoxy to the inside of the head gap, and push the handle through the gap, by hitting it with your hammer. Do this until you have a bit of the handle projecting out of the hammer head on the other side.

6. Wait for 24 hours to let the epoxy cure.

Making a hammer can be a strenuous project due to the massive amount of hammering that has to be done in order to get the hammer head into a good shape.

[BONUS 2/3] A DISCUSSION ABOUT DAMASCUS

So what is Damascus? It is an alloy of steel, made using two or more kinds of steels. It has greater toughness and hardness than normal steel. Damascus has a legendary status among bladesmiths. This chapter is a discussion about Damascus steel and it also contains several points regarding forging Damascus steel.

As stated briefly in the introduction, when the Spanish invaders invaded the city of Damascus, their swords were outmatched by the ones used by the defending army. The Damascus swords could literally pierce through the invaders' swords.

Sometime in the 1700s, the supply of Wootz steel to Syria stopped. This lead to the death of the original Damascus swords. The original method of making Damascus steel was also lost. The modern method to produce Damascus is known as billet welding. To know more about billet welding, read the question and answer section below.

Note: This is not a step-by-step guide like the rest of the book.

Here are the answers to some questions commonly asked about Damascus steel:

1. How is forging Damascus steel different from forging normal steel?: Damascus is the mixture of steel and one or more metals. It is made by stacking multiple strips of steel and metal over each other and welding over it. The block that is created is known as a billet. The billet is then beaten heated and beaten into a single piece of metal. After this the process becomes similar to making steel knives.

2. Why is Damascus so popular?: Damascus was quite extraordinary for its time. It had a legendary status due to it being extremely flexible but hard at the same time. This made the blade virtually unbreakable yet it was able to hold an edge for a long time. It has made quite the impact.

3. Is making knives out of Damascus solely suitable for advanced bladesmiths?: For the absolute beginner, I would recommend that you first make several knives out of normal steel, and then experiment with Damascus. Typically it is harder to machine than a steel like 1084 steel. Nevertheless, do try making something out of Damascus eventually!

A Damascus knife is a prized knife for any bladesmith. The knives made out of Damascus steels fetch way more money than the ones made out of normal steel. They are also famous for the wavy patterns that they exhibit.

Some small details about Damascus steel:

1. Since it is harder, Damascus steel is harder to machine.

2. Damascus steel can be etched using acid. In this method, the acid eats away at both the steels, but at different rates. As a result one of the metal stands out more than the other, causing patterns to come out in knives.

3. Damascus steel has its roots in Ancient India. Through the use of Wootz steel and a refining technique, they were able to make purer steel with a higher carbon content, thus making it exceptionally hard and capable of holding an edge for a long time.

4. Damascus steel is also more suitable for longer blades like swords, because of the flexibility that they provide.

5. Another unique characteristic of this steel is the presence of exceptionally high carbon content in the steel. The steel actually contained carbon 'nanotubes'. They are a type of carbon structure that is extremely light yet hold a large weight, and are being developed by modern scientists and armies.

6. Try and weld darker steels like the 1084 steel with brighter ones to create patterns with better contrast.

[BONUS 3/3] A PEEK INTO SWORDSMITHING

The natural progression from making knives is to make swords. This bonus section discusses swordsmithing briefly. Before we think about smithing, let us look at the different parts of a sword:

Blade: This is the cutting part of the sword. But sometimes the whole sword is also referred to as the blade. The blade of a sword has two edges.

Handle: The part of the knife that you hold. It is generally made out of wood.

Choil: A small indent in front of the edge, which marks the end of the edge.

Tang: The part of the blade inside the handle.

Cheek: The flat part of the blade.

Bevel: The slope towards the edge of the sword.

Pommel: It is the weighty end of the sword handle. It is usually circular in shape.

Guard: It is the part that separates the handle from the blade of the sword. It is commonly found in a rectangle form.

Hilt: It is the area of the sword consisting of the guard, handle and pommel.

HOW IS FORGING A KNIFE DIFFERENT FROM FORGING A SWORD?

There are some small but fundamental differences between forging a knife and forging a sword:

1. The sword is a combat-focused weapon. A knife can be categorized more as a tool cum weapon.

2. A sword has to be flexible, while being able hold an edge well. A knife has to do the same things, but flexibility is a much larger issue for swords due to their size.

3. They require different steel for smithing. While a higher carbon steel like 1095 steel is recommended for knife making, it is not suitable for absolute beginners when it comes to swordsmithing.

4. Generally, a knife has an edge on only one side. Whereas a sword is usually double-edged.

5. Swords are much heavier than a knife, due to the size of the blade. To offset the weight of the blade, a pommel is used on the handle which helps in balancing weight.

Frankly other than these factors, knives and swords are pretty similar to each other!

This was a bonus preview of my upcoming book, *'Intermediate Guide to Bladesmithing'*.

'Intermediate Guide to Bladesmithing' is a book made for the aspiring bladesmith who wants to take his craft to the **next level**. Its main focus is on swordsmithing and the smithing of Damascus. Consider it as an extension of this book.

Swords are one of the most elegant weapons, ranging from the normal steel sword to the legendary Damascus swords and Japanese katanas. Making such elegant weapons requires a skilled bladesmith. This book is geared towards giving you the steps to attain such skill. Swordsmithing in many cultures is a very prestigious and honorable craft. By reading this book you are taking your first steps in this wonderful profession.

CONCLUSION

Thank you for taking action and reading this book!

I hope this book was able to guide you towards the start of your bladesmithing journey, with the content in this book. The next step is to master the processes given in this book and to move on to sword-making!

Finally, if you enjoyed this book, then I'd like to ask you for a favor, would you be kind enough to **leave a review** for this book on Amazon? It'd be greatly appreciated!

Bonus: Now that you have made a knife that you are proud of, wouldn't it be amazing if you could sell it too?

Get my knife selling secrets with my **100% FREE** e-book, *'Bladesmith's Guide to Selling Knives'*.

All you have to do is go to **http://eepurl.com/gbq6Sb,** and enter your email and name.

Thank you and good luck!

Intermediate Guide to Bladesmithing

Make Knives, Swords and Forge Damascus

Wes Sander

© **Copyright 2018**

All rights reserved.

The content contained within this book may not be reproduced, duplicated, or transmitted without direct written permission from the author or the publisher.

Under no circumstances will any blame or legal responsibility be held against the publisher or author for any damages, reparation, or monetary loss due to the information contained within this book, either directly or indirectly.

Legal Notice

This book is copyright protected. This book is only for personal use. You cannot amend, distribute, sell, use, quote, or paraphrase any part of the content within this book without the consent of the author or publisher.

Disclaimer Notice

Please note the information contained within this document is for educational and entertainment purposes only. All effort has been executed to present accurate, up-to-date, and reliable complete information. No warranties of any kind are declared or implied. Readers acknowledge that the author is not engaging in the rendering of legal, financial, medical, or professional advice. The content within this book has been derived from various sources. Please consult a licensed professional before attempting any techniques outlined in this book.

By reading this document, the reader agrees that under no circumstances is the author responsible for any losses, direct or indirect, which are incurred as a result of the use of information contained within this document, including, but not limited to, errors, omissions, or inaccuracies.

INTRODUCTION

Before We Begin

Firstly, ask yourself, "Why do I want to make a sword or knife?"

The next question should be, "What do I plan to *do* with the knife or sword?"

Now, that question may be perceived as strange by most people. Why should I make a knife when the market is filled with a lot of low prices, yet very attractive blades, for sale?

I've been asked this question a number of times from quite a lot of people. Some people are okay with a small knife that can be carried in their pocket to use in cleaning their fingernails or opening an envelope, etc.

Others want a knife that can be used in hunting for cutting up game after a successful catch. Some see the knife as an important tool, one to always care for and not be without. These people desire a knife that can last for a lifetime, and that will hold an edge.

A large number of people who make knives, or are thinking of making knives, are doing so because they are not satisfied with the knives available in the market, and making one is the only way they can get their desire.

Knives and swords reek of adventure, gallantry, swashbuckling, and gallantry. Their creation is full of myth and mystery.

I am presenting techniques and guidelines for making knives, swords, and Damascus steel with this book. You can utilize this book to make one or numerous knives or swords, or it can be used as a handbook to start up a workshop and eventually earn a living with the craft.

Let me make a suggestion that you first read the whole book from cover to cover as though it were a novel. Once you are acquainted with the whole range of consideration, you can start to realize your knife-making goals in a step-by-step manner.

Let's get right into it!

CHAPTER 1: COMPREHENSIVE GUIDE TO KNIFE MAKING

Despite the fact that we use knives daily, at least in the kitchen, have you ever thought if it would be possible for the people of early civilization to survive without a knife? Today, various types of knives are used, such as pocket knives, camper and hunting knives, paper knives, bread knives, etc. A knife made of flint was a source of food and was used to provide shelter in the olden days. Some people say the invention of the wheel was the greatest invention to have happened, but how was the wheel built without cutting equipment? The first hand-making equipment that was used to control the world around us was the knife.

Different materials are used to make knives. Earlier, it was bronze, flint, and copper, then lastly, it was steel. With the modern equipment used in knife making, the methods for making knives have also advanced.

Folding knives and knives with fixed blades still exist. So what has really changed over the years? The materials are what have changed the most! Recently, we use sophisticated materials to make knives, such as titanium, steel ceramics, or carbon.

Hundreds of years later, a knife became more than an object for plant cultivation and more than a weapon for hunting animals. We use a knife every day because it has found its way to our tables. Knife making became a real art. Knife engraving, adding gemstones and customization are now common.

Every knife that you make can be a real stunner and a show stopper. It will astonish you of the magnificence and the distinctness of the knives you make. They are elegant and sleek, and you can create each with your own touch. An advantage of making your personal knife is that you can choose the steel that meets your prerequisites.

What can we call a Good Knife?

Before going any further, we have to go back and take a look at the entire concept of people and knives. Just as with other things, there must be the same number of value structures and attitudes as regards knives as there are people. This chapter can therefore only be of the utmost value if you make use of the information in it as it suits your desire.

Of course, knives can be anything from an exquisite work of art to an item in a shoddy dime-store to a powerful weapon or tool. How amazing a knife is, is solely dependent upon the user's value for it.

A knife's value to you is based on how good it satisfies your desire. You may like a knife that you bought ten years ago so much that its quality hasn't waned. On the other hand, you may dislike a "good" knife that is quite heavy to carry or doesn't hold an edge as much as you desire. Or you might just prefer a cutting-edge, corrosion-resistant, high-performance, computer-age model over a simple knife with an old wooden handle. And, apparently, another knife lover could just love it the other way around.

The fact of the matter is that you are the ultimate judge, either as the user or knifemaker.

Design

It's not enough to say that a knife should look fantastic and feel comfortable to you, it should also be strong enough and not overly heavy. "Overweight" is one of the most annoying faults of most knives, especially those designed to be carried in a sheath or in the pocket.

The knife's design should be such that the flow, of weight, mass, and lines, as well as the material's physical strength and the edge's keenness, should all work together as an energy "transmitter," from the user's hand to the cutting edge. The knife's design is a physical expression of the purpose it was intended for.

Knife-Making Steel

We need to explain some things about the materials you are going to use before we proceed to knife-making equipment. You should focus on many things when choosing a knife making steel, especially if it's going to be your first time making a knife.

Knife makers use several kinds of steel. You should know some basic information about those that are majorly used. Depending on how you make knives, either by forging or stock removal, there are a number of steel types to choose from.

All steel needs to be treated with heat. As a starter, you might not have a heating oven yet, so go for something simple, something you can easily heat-treat and work with. Stainless steel, carbon steel, and tools steel are the commonly used steel in knife making. There are plenty of steel comparisons on the internet that will help you choose. The following are the steels most commonly found in knife blades.

O1 steel: It is perfect for beginners. It is very tough, oil quenched, easy to work with, wear-resistant, and holds an edge perfectly. It needs to be taken care of to prevent rust.

1084: It is also good for beginner knife makers. It is good for uncertain heat-treating. It can be purchased anywhere. It does not need to be soaked for long.

1075: It is the same as above and quite affordable. It allows heat treatment with a blowtorch (not so high temperature for heat treatment, 830 degrees or thereabouts). This makes it a perfect choice for beginners.

1095: Percentage of carbon is high (95 percent), and it is mostly used for making knives by forging. In my opinion, it is the best knife making steel.

Stainless steel: It is more expensive. The blade will not rust, so there is no need for maintenance. There might be a need for someone to heat-treat it for you. It sharpens easily.

Top Tools for in Making Knives

If you are just starting and you don't intend to take knife making serious, do not waste too much money on the equipment. Buy equipment that is not expensive and is important. Simple equipment is required for simple projects. If you decide to make a knife using the stock removal method, this means all the gear used when working with fire (big hammers oil and anvils) are not needed.

Tool #1: File

This is the favorite equipment of many bladesmiths—good old-fashioned file. Great things can be done with this simple equipment (of course, except drilling a hole). Files are the first equipment you need to get, and they are not expensive at all. Smoothing, grinding, finishing—all can be done with files that come in various shapes and various grit sizes. Many people made their first blades out of an old file. Files can be bought at every metal shop, and they cost no more than $10. You still need to have a set of files in your garage, though there is a lot of power equipment that can replace the files and do the job much faster. They also require manual to work.

Tool #2: Clamps

When you are working, you need something to hold the knife. So buy several pairs of clamps; you will need them. They are cheap, so try to buy various kinds if possible. You can get C-clamps and welders for starters, then progress to bar clamps, one-handed clamps, pipe clamps, and so on. You can purchase them online at Home Depot at $10 or more depending on the type.

Tool #3: The Hacksaw

Also, an important piece of equipment with a high-quality blade is the hacksaw, which is needed especially in the steel-cutting and steel-shaping process. The hacksaw is always needed to start the knife-making process. The job can be done faster with power equipment, but they can't be used in tight corners. You can order a hacksaw at Amazon starting at $20 and make sure spare-blade kit is ordered as well.

Tool #4: The Bench Vise

Everyone who deals with metal requires one bench vise at least. When buying a bench vise, there are a lot of options to choose from, so simply start with the one that enables you to change the orientation of your work: a 360-degree swivel base adjustment. A good bench vise can be found for $100, but you can buy a used one if you feel that is expensive. An all-around bench vise size is generally five inches, but you can opt for something much bigger.

Jaw caps should be mentioned when we talk of a bench vise. They are designed for the protection of your knife. They are normally made of rubber, plastic, leather, aluminum, or copper; and they can go for around $20 on eBay.

Tool #5: Grinding Abrasives

When it comes to polishing a blade, fine grinding and finishing works, Cones, sandpaper, rubber abrasive wheels, points, and sticks are all put to use. If you're a knife maker and you want to give a unique final touch to your work, then a rubber abrasive is a *must-have*.

Tool #6: Drill

A drill press is very convenient, and it is a better choice, while hand drill with some bits for drilling steel is very good for a beginner knife maker. You can consider buying a used drill press with a drill vise. It will save you time by improving your accuracy. A hand drill can be bought for about $30. A ten-inch drill press at Sears costs around $121. A set of Cobalt drill bits costs around $30 on eBay. A four-inch drill press vise costs $17, and it will get the job done very fine.

Tool #7: Sharpening Stone

In the knife-making process, the last step is sharpening. When it comes to buying a sharpening stone, there are numerous options to choose from. The price ranges between $20 and $200, even higher sometimes. A lot of brands, stone sizes, grit textures, and stone shapes exist in the market, and the material of the stone is a determinant factor for the price. Most times, knife makers make use of diamond sharpening stones, but you can try water stones, Arkansas stone, ceramic stones, or whatever suits you.

Tool #8: Safety Gear

When grinding, cutting metals, and heat-treating steel, make sure you always use safety equipment: dust masks or respirators, gloves, and safety glasses. Serious injuries can be prevented by safety equipment; it can also protect your hands, eyes, and lungs from heated metal hazard dust particles. Gloves and glasses are very cheap, and instead of a dusk mask, a respirator is always a better solution. A high-quality 3M respirator can be gotten at Home Depot for $100.

A fire extinguisher is required for knife makers who make knives by forging steel. A five-pound rechargeable fire extinguisher doesn't cost up to $50. This is a small investment you can make to prevent fire outbreak at your workplace.

Tool #9: Dremel or Another Rotary Tool

Another tool is Dremel. Although it's not an important piece of equipment, it can be used in cutting materials, cleaning rust, grinding details, jeweling, or customizing with mounted abrasive cones, cut-off wheels or small rubber, points, and bits. For most knife makers, Dremel 400 is the first choice, and you can find it online for $77. You can check eBay to find the best offer. You can buy a flex shaft attachment or a handpiece along with a Dremel, which is a very convenient accessory since the motor unit won't have to be held in your hand.

More Advanced Pieces of Equipment

You will need one of the aforementioned equipment at a time or the other. People who want to do knife making for leisure can invest less than $1,000 and have a very good basis. Although you might have to invest more if you are in between, moving from doing knife making leisurely to being a full-time knife maker. In case you're looking at becoming an expert at knife making, discussed below is the equipment you will need to have in your garage or shop.

Tool #10: Belt Grinder

We are talking about 2"×72" industrial knife belt sender made for professionals, not the small bench belt grinders. Purchasing a smaller 1"×42" (belt dimensions) belt grinder shouldn't be a choice here. If the equipment is too expensive for you presently, don't buy it yet. Do not purchase a grinder that won't get the job done.

Belt grinder prices vary a lot depending on many factors—motor speed (variable speed control is a fantastic option to have), wheel speed, wheel size (usually eight inches, though you can find any dimension you want), motor power (if you intend to grind long steel bars and tick, don't go 1 HP), accessories, body material, tracking adjustments, ability to quickly change belt (quick-release mechanism), and so on. There are similar specifications in all professional and powerful belt grinders.

An affordable belt grinder for knife making, including shipping, costs about $600. High-end models can even be as high as $3,000. Check the internet to find a used one that's still in a good condition.

Tool #11: Heat-Treating Oven

As said earlier, steel has to be given heat treatment. In the knife-making process, heat treating is the most essential, and demanding shops that sell steel usually offer heat-treating services also, so heat treatment can be outsourced. However, you'll need to own a heat-treating oven (some call them kilns or heat-treating furnaces) if you want to take knife making serious.

You can surely use a torch to heat a two-inch stainless-steel blade, but trying to heat a five-inch or a longer blade at the constant temperature of 1,000 degrees can be quite complicated. While the other parts of the blade remain cool, you can have the correct temperature at the center of a blade. One of the reasons you should buy a treat oven is because irregularity in hardness can lead to failure as blade simply won't hold the edge.

When you want to purchase a heat-treating furnace, what is the most essential thing to know? The oven's major function is to heat the blade to an accurate high temperature and to maintain a constant temperature. This means it must have a manual or digital temperature controls and a dependable temperature controller. The capacity of an oven depends on the blade you want to heat. The maximum temperature it can attain is also essential, and the temperature standards are 2,000 °F and 2,350 °F.

The oven is an expensive piece of equipment, and its price depends on its size. An oven that has a chamber size of 6.5" W × 4.25" H × 18" D costs about $1,200.

You should also buy industrial stainless-steel foil and ceramic racks. You will need a heat-treating foil to protect the blades and prevent them from discoloration and scaling; you will also need a rack to position the blades in the oven.

Note that heat-treating ovens are mostly electric, and they use up a lot of electricity.

Pro Tip: Tool #12

Pizza? No. Beer? No. AC/DC soundtrack to enjoy while grinding? No.

Tool #12 is YouTube, Instagram, or some other social media network! Publicize your work, your current project, and have a great time. Seek opinions and advice. Like or follow other knife makers. Endeavor to learn something different, something new. Copy some ideas and improve on them. Share your knowledge, learn some new knife-making skills, and find out about new tools. Give valuable information to people who are in need of new handle materials, searching for heat-treating services, or probably just need assistance selecting the best steel for their project.

As you work, record videos and upload them on YouTube to help people pick up new ideas from you. Join a community of large knife makers and have a positive influence on upcoming and already-existing knife makers.

Step-by-Step Guide

Step #1: Sketch the Knife

In getting started with making a knife, sketch the design and shape of the blade on the paper. Ensure that you keep it at a scale size of 1:1 for easy construction. The dimension of the blade is dependent on your choice. However, large knives require a lot of steel and can be heavy.

Step #2: Choose Tools and Knife-Making Steel

You should stay away from stainless steel because it has to be subzero-tempered and isn't a good choice for making good blades. An excellent knife-making steel is the 1/8-inch-thick carbon steel (01). It is a good choice for producing blades because it is easy for drenching.

When making handles, wood materials are a perfect choice. Although you can use any material you want to make the handle, the materials that you can combine with rivets are G10, micarta, and kirinite when making the full-tang knife. These materials are the best to make knives because they are waterproof.

Use a permanent marker to trace the blade onto the slab. This will serve as a guide when cutting the steel. Ensure to trace the tang accurately specifically because the blade and the tang are connected as one piece.

You can use a hacksaw, bandsaw, vise, an angle grinder with cutoff wheel, drill, grinder, and safety wear.

Step #3: Cutting the Steel

Cut a rectangle around the traced blade using a hacksaw. This is to separate it from the main slab. To cut a thicker steel, make use of a stiffer hacksaw, then grind the rectangle down to ensure it forms the blade profile.

Put the rough-cut blade in a vise and grind out excess steel. Follow the guidelines when molding the profile. Complete the blade shape using the grinder. Then use the grinding wheels to softly grind the edge into a slope. Make this slope on both sides of the blade to get the desired blade edge.

Afterward, use a drill bit of similar size as the rivets you want to use. Place the holes in the tang. You may need a different number of holes depending on the size of the blade.

Sand the blade at this point with sandpaper or finer grits. Ensure you sand every surface of the blade, including every scratch. This is to help improve the blade's shine and quality.

Step #4: Heat-Treating the Blade

The perfect way is to use a forge in heating the blade. A forge is a type of brick- or mud-lined fireplace used to heat metals. A gas or coal forge can be used for this procedure. A torch can be used for smaller blades. A torch is a thick stick with materials burning on it; it is used as a source of light.

At this stage, the blade is ready for quench hardening. Quench hardening is a mechanical process for hardening and strengthening of steel. To cool the blade, you must douse it in a hardening bath, usually a bucket of motor oil. Immerse the entire blade in the bucket.

Leave the steel in the fire until it gets an orange glow. Place it close to a magnet to be sure it is sufficiently hot. Once the steel attains its optimum temp, it loses its magnetic properties. If it doesn't stick to it, let it cool sufficiently at ambient temperature.

This can be repeated three times. On the fourth time, do not cool the steel in air; rather, dip it in oil. This requires extra care because there will be fire eruption when the blade is dipped into oil; therefore, cover the exposed parts of the body using personal safety gadgets. When the blade is hardened, it could break easily if it is dropped; thus, handle carefully.

At this point, set the oven (thermal chamber used for heating) at 420° Celsius. The blade should be placed on the center rack and heated up for sixty minutes. Once the 60 minutes is over, the heat procedure is finished.

As before, sand the blade with smoother grits of paper and clean the blade for some extra shine.

Step #5: Making a Knife Handle

There are two sections of a handle for a full-tang knife, one on each side. The wooden piece should be cut using a hacksaw while the pieces are sand in simultaneously to ensure that the two sides are symmetrical.

Once that is done, drill the holes on each side for bolts. Attach a vise and let it dry overnight. A saw is needed for finished products and handle modification. Insert the bolts, leave about 0.125 inches, and peen them using a ball-peen hammer. Then you can file them down. Bear in mind that you must also sand the handles.

Step #6 Blade Sharpening

You will need a large sharpening stone for this process. Pour sharpening oil on the jagged side of the stone. Then bevel the blade at an angle of 20 degrees from the zone of the sharpening stone. Move the blade against the stone in a cutting movement.

The handle should be drawn up while moving the blade to sharpen from bottom to the top. After a couple of moves, switch the blade over to the other side. When you have made a sharp edge on all sides of the blade, take it to a fine stone.

Step #7: Try Out the Knife

Cut a few pieces of paper with the knife. A well-sharpened blade should easily cut the paper into strips. Once the knife performs well, congratulations! You've got your DIY knife.

CHAPTER 2: COMPREHENSIVE GUIDE TO SWORD-MAKING

Have you ever thought of making your own sword? I don't simply mean a knife or blade. I mean a real, full-working, and harmful sword. Most of the sword makers use old techniques sharpened by the ages. However, some techniques are more recent. You will learn a couple of new things during in this section.

Caution: The making of a sword is quite easy for someone acquainted with their hands, yet it takes patience and requires full concentration. The task ahead is daunting, and you would need to invest hours thinking, wandering, and endeavoring to solve the next stage the most ideal way. In any case, with this section, it becomes easier, and anyone can make a sword in a significantly auspicious way.

Also, most steps along the line can be done in other ways depending on the alternative tools you have.

Step 1: Materials

Full blacksmithing tools comprising of the following:

Required tools: forge, tongs, hammer, and anvil

Shaping tools: bench grinder, belt grinder, angle grinder, and cutting torch

Steel-finishing tools: belt grinder, angle grinder, belt sander, drill bits, drill squeeze, wire wheel, a variety of sandpaper, and steel files

Woodworking tools: drawknife, saw, files, table/hand sandpaper, linseed oil, and multifunction knife

Optional yet recommended: induction heater, power hammer, and air hammer

Be cautious. Ensure you're skilled with each of the tools you use.

Step 2: Dreaming and Designing

At the initial stage, you need to understand what you need your sword to look like once completed. It is imperative to have a good plan before commencing. Keeping the design simple makes the work much easier and better. Stick to a straightforward geometric design rather than complex fantasy-like thoughts. At first, sketch a design on paper and try it out on cardboard or something stronger.

Gathering ideas from history is a smart idea. Carry out some research on what they actually look like in the past. See what current swordsmiths are producing too.

Step 3: Finding the Steel

This is a crucial aspect. The steel used for a sword is really important. The use of high-carbon steels is strongly recommended. Few steels can be hardened. Unless you only plan on having the piece as a beautification, you need to choose a decent piece of steel.

You can determine what sort of steel it is, depending on the sparkles produced. High-carbon steel will produce sparkle that splits off into a few branches.

If you don't have a forge, then you need to make one. If not, get a piece of steel that is almost the same size as the sword and move past the forging process.

Step 4: Starting Off

To start forging, get the steel to the correct size and shape. Cut off the extra parts of the piece if the tip of the steel tapers is more than you need. It's better to cut the opposite side before the hole in the steel.

If you have the right length, simply cut the tip for the tang. Use a cutting torch to cut the sides off. You could make it about ten inches in length.

Now that you have your right steel measure, it is time for the fire!

Step 5: Shaping into "Your" Blade

If the size of the blade is different from your ideal blade style (i.e., the blade is excessively wide and not sufficiently long), you should decrease the width of the piece. This all can be carried out by hand, but a power hammer might probably be needed depending on the steel's mass.

Heat the whole section with a burner gas forge till you see a yellow glow. With the hammer, carefully apply blows continuously along the piece length.

After a few heats, it will start to increase in length and thickness while it decreases in width. In the process, the sides of the steel will start to thicken up more than the center. To balance this, start hammering it on the blade's flat side.

The entire procedure requires a lot of forward and backward movement. By using hand tools, it is a lot less demanding with limited damage possibility.

The general goal is to extend the blade by compressing the width and reducing the resulting thickness.

Change to a pneumatic hammer for better work. Hit the two sides of the blade to make it appear more like a sword.

Once the appropriate width, length, and thickness are attained, start squaring up the piece as a whole. The aim is to get the sides parallel, fit in the bends at the tip, and attain a uniform thickness of the blade.

Now, the blade has its actual appearance and is set for the next step.

Step 6: Grinding the Appropriate Profile

When you've accomplished a uniform thickness on the steel, the following stage involves grinding it to the ideal proportion.

The initial step is to imagine what you need the blade to look like. Ensure the regions you mark out have thick-enough steel. Include this in the planning procedure. You should make certain that the blade has enough material.

Commence profiling with an angle grinder using a thin cutoff plate. Get so close to the real shape without damaging the blade. It is better to leave abundant material and do the finishing with a grinder.

You can design the tang in different ways, yet it has to follow the simple idea of curving the two sides with a tapering tang.

Once you've used the angle grinder, finish up the profile either using a belt grinder/sander, a bench grinder, or an angle grinder with a grinding wheel.

Before grinding, all the slag on the steel must be removed because it dulls the belts rapidly.

To remove the surface slag, a variety of things can be used:

- wire cup on an angle grinder
- flap sander on an angle grinder
- wire wheel on a bench grinder

The use of a wire wheel is recommended except it could turn out to be way too strong. If that is the situation, then use the cup brush or the flap disc. Try to just remove only the slag and secure the steel.

Step 7: Making the Bevels

At last, the steel now takes the form of a sword. The next challenge is to determine what you want the cross section of your blade to look like.

The cross-section determines what the blade would look like if it were to be cut in bilaterally on the blade's perpendicular axis.

The primary concern when preparing your cross section is to consider the sort of grind you need. The three major types of grind are flat, concave, and convex. These grinds all have their own technique, advantages, and disadvantages.

- Flat grinds are straightforward as they sound. The grind is an angle across the whole blade's surface. It is blunter than a concave grind because it has more material. This grind can be efficiently sharpened in the "field." Though it is a strong style, it is not the hardest.

- Concave grinds bend inward toward the cutting edge. They are the sharpest of all grinds, with the best cutting edge. However, they also have the weakest edge. Such factors should be put into consideration. The concave grind is the hardest of all grinds to make.

- Convex grinds are generally found on axes. The edges curve smoothly. This provides the blade with the best strength of the three grinds. However, a convex grind will give you the bluntest edge with the lowest cutting force. This style is considered the least demanding to cut.

The other part of the cross section is the fuller, sometimes inaccurately called a blood groove. The fuller gives a lighter weight without sacrificing quality and strength. Its properties are almost the same as that of a construction I-shaft.

You may decide to go with the convex grind if you can't maintain consistency across such a wide surface. Try not to make any fullers if the blade you made is way too thin.

At the start of the grind, first, ensure the blade is well hammered and ground perfectly flat. If you are satisfied with it, you can move on to a bench grinder or belt grinder. Begin by slicing off material at about an angle of 45 degrees. Make the grinds using a dull belt, then gradually move up to a sharp one. If you begin with a sharp belt on a corner, the belt will dull rapidly.

Cutting bevels involves going gradually at a time. Allow the machine to do the work. Take your time while working on any project. This stage will take a while. Utilize the unsupported portion of the belts. Consistent pressure should be applied at the same angle. Start from the shallow ends. Afterward, let the machine work its way to the center of the steel's thickness. Cut the bevels to the same proportion from both sides till less than about 1/16 of an inch is remaining.

Step 8: Heat Treatment

This is the most important step. This transforms a sharpened shard of metal into a splendid weapon. If you have no idea what kind of metal it is, oil is the safe choice to use. Quenching in water can make certain steels crack, and this is the reason for it to be avoided.

Steel quenching hardens steel because it transforms carbon molecules into a tight lattice arrangement by quickly cooling in something like oil after they are heated up. The steel needs to attain its optimum temperature, which can differ depending on what sort of steel you have. Again, the convenient step is to simply go past the magnetic point of the steel.

When heating the piece up, be careful not to overheat it. Begin with a low temperature and gradually raise it. Ensure to evenly heat the blade, concentrating on the tip and edges. The tang is not a vital point, but its joint is.

Use a piece of three-inch PVC pipe to hold the oil, then glue the end cap on and clamp it to a firm table. Ensure the section of pipe is a bit longer than the sword. Ensure the oil is warm before quenching. This can be accomplished effectively by heating up a piece of metal to orange and placing it in the oil as this will help warm up the oil.

When the blade is uniformly heated and the oil is set to quench the blade, dip the blade into the oil and forcefully swirl it at the same time. This ensures an even cool-off. Allow about 30 seconds for the blade to stay in the oil, followed by air cooling. When cooled, transfer it to a very high grit belt to quickly eliminate the scales.

Tempering comes after quenching. The quenching makes the steel fragile, and tempering brings back the sturdiness of the metal. Both the tempering and pre-quench temperature, as well as time and liquid type, will be determined by the type of metal. Evaluate what you have so as to find the best possible way.

Step 9: Blade Finalization

This stage is all up to the maker's choice. You could choose to use a Scotch-Brite belt to clean it up.

First, get the blade to the desired shape with 60–80 grit. Then move to 120 grit. Ensure that all the deep belt scratches are eliminated with each increase of grit. Keep on expanding the grit to 220. Then 400 if you wish.

Once you've removed all the deep scratches with the 120, most bladesmiths will move to hand-sanding. The process takes a while, but the outcome is worth it. Go on to the following grit, perhaps 220. Repeat the procedure and sand until only the 220 grit sandpaper scratches are left. Sand simply with a little piece and pressure by hand. You can decide to wrap a piece around a wood block. Another choice is to use an orbital sander. Be cautious though. Apply moderate speeds and dull paper. Be cautious of the sword edge. When sanding one side, apply more pressure so the sharp edge is pressed to the table and can't cut your finger as you sand. Next, move to 400 grit, then 800, 1,200, and possibly 2,000. The general principle is to double the number as you go up.

Patience is important.

Step 10: Pommel Creation

Another important part of the sword is the cross guard and pommel. These can be made out of bronze or brass.

The role of the pommel of a sword is to balance the weight of the sword and to aid in the mobility of the weapon. Without it, the sword will be very stressful to wield. The importance of a pommel cannot be overemphasized as its usefulness is backed by history.

Get the metal from a pool heat store. Pools use crude synthetic chemicals, so the metal must have the capacity to withstand corrosion; therefore, they use materials like copper or brass. To plan for the melting, clean the muck off the pieces after some time, with lots of hitting and smashing.

Use a crucible with a gas burner in a forge in melting. Preheat the pieces at the top of the forge, not inside. The aim of this is to aid in the melting and to remove more of the impurities from the metal. Every time the piece melts down, clean off the slag from the top and add another. As soon as the crucible is full, it is time to pour.

Empty this into a section of pipe. Place the pipe into some sand kept in a bucket and do the pouring. The leftover material could be made into a mold for bars for melting ease later. Allow time for it to cool and cut out the piece from its pipe.

If you have an appropriately sized piece to begin with, you could skip this prior steps for the pommel. Any pommel design can be accepted. Numerous options exist out there. Put the piece on a lathe after you have obtained your "stock" piece.

The pommel aids in holding the whole sword together. After some time, weld some threaded rod onto the tang, which then runs through a hole in the pommel. This is the initial step. Place a drill bit on the lathe (the rod's size) and drill it through the pommel. Afterward, drill a hole for the nut to join the rod to the pommel so it could be concealed. A regular drill should be used to get a large portion of the material out, followed by an end bit to ensure it is square on the bottom side. With that, proceed to the cross guard.

Step 11: Cross Guard

The cross guard has a couple of vital functions. Firstly, prevention of an enemy's sword from sliding down your blade into your hand. In the absence of this, a blade will clash, slice, and cling. The other vital part is to prevent your own hand from going up into the blade, which is not likely to happen but possible.

Any material can be used for this, just like it is with the pommel. A piece of bronze, brass, steel, titanium, aluminum, etc. could be used. Anything can be utilized. You can take any design with it also. Square or rounded is great. Hello, be creative. Anything can be used. Create a dragon. Okay back to the real world.

Get a piece of material, brass or bronze. Their alloys can be extremely challenging to work with. If it gets excessively hot, the material will disintegrate like a failing sandcastle. Some individuals use the term *red short* to describe the phenomenon. Also, the material has the tendency to work-harden, implying that by bending and working it, it gets more difficult to move. It must, therefore, be strengthened by gradual heating and cooling. It tends to crack along failure points even if it doesn't disintegrate. Working it cold is simply the solution to this. To smash it into shape, power hammer should be used. This is the most recommended from experience.

Once you have a piece that's large enough, drill some holes in it to allow the tang to go through, and file it to make it clean. It must be sized so that it lay flush with an appropriate slot in the piece, then set a bar through it. The bar will at that point set into the vertical milling machine vise. You will be able to make its sides totally parallel with the tang's whole/line. After that is accomplished, go to the grinder to clean the piece. The same way of cleaning and wire wheel as before.

Connect it to the sword by soldering it with unadulterated lead solder and flux. Ensure you fill in every one of the gaps for strength.

Step 12: Wooden Handle

The wooden handle is next. For the tang to go through, the handle has to have a hole. This is done specifically with both drilling and burning. A straight hole is set for the tang by the drill, making a lesser material needs to be burned. Afterward, a mock tang is built to burn a size almost the same. Any extra space after cleaning is done will be filled with epoxy when it is glued. Now for the exact procedure.

The initial step was the drilling of the hole. Try not to burn out the hole, because it is quite difficult and may end up unsuccessful. The best way to go is to drill a hole. Place the piece in the vertical milling machine and drill a hole smaller in size than the tang width. A drill press can also be used just as effectively by hand, needing a larger stock piece, and leaving room for more error. Begin with a typical length drill bit. After the small-sized bit, utilize a longer one that can now fit in the machine and can be set into the wood due to the hole. You could drill the remaining by hand if you are still not fine with the hole piece after that.

With the hole now drilled, proceed to the grinder and clean off all the corners, rounding the piece. With it more cylindrical, place it into a lathe and begin to shape the piece. After sizing down, sand the lathe. With the handle now finished, it's time to do the pommel connection.

Step 13: Pommel Connection

Add some threaded rod to the back to connect the pommel to the sword. Get the right-sized bolt to go through the pommel and cut the head off, leaving you with threads. After this, cut a groove in the thread to slide onto the tang.

You now have the wooden handle's size. So to match, cut the tang of the sword. Weld the rod on and try to slide the handle on too. Grind the weld down if it is too big. After this, slide the wooden handle on, followed by the pommel. If you observe that the handle didn't line up square with the cross guard and pommel, shape the top and base of the handle to match.

If the connection satisfies you, slide it all together and place a nut and washer into the pommel.

Step 14: Gluing

After all the pieces have been prepared, all that is left to do is to glue them into a full sword. A two-part epoxy should be used as a bonding agent.

With everything arranged, mix the glue. Every part needs glue. Slather it throughout the tang and pour into the wooden handle. Also, line the top and base of the handle. The handle should be slid onto the tang and the pommel. Slide on the washer and crank the nut down.

Leave the glue to dry and get cured for several hours.

Step 15: Polishing/Buffing

Buff the sword using the glue set. An angle grinder with a fabric wheel and a buffing compound should be used. Clean the whole cross guard, pommel, and sword blade. Before wrapping it in leather, sand the handle and apply linseed oil.

Step 16: Leather Wrap

Should you not really like the way the wood turned out, wrap the sword's handle. This step isn't compulsory though.

Firstly, locate a section of leather that conveniently has the correct length. Wrap it around the handle and take note of the spot on the top and base of the wood of the leather by marking it. Draw a demarcation just between the marks and cut it with scissors.

For stitching, punching is first required because leather is extremely tough. Before punching, mark the spots where you have to punch. A fork could be used to press down to make the mark. Do this till you've marked all the points. From the sides, ensure the marks are about an eighth of an inch.

Using only a punch and a hammer, punch the holes on a scrap piece of wood. An awl could also be used. Drilling with a very small drill bit is another common technique for thick leather. Punch all of the holes, and then it's ready for stitching.

Begin the stitching process from the handle top by looping it underneath. Ensure it is tied at that point and wrapped over. Stitch it in a similar pattern down the handle. Rethread the line repeatedly through the same spot at the base to ensure that it stays.

Using two needles beginning at the top and acting like you're tying a shoe is a better approach to this. It can be finished using a knot at the end.

Step 17: It Is Done

There you have it finished. A step-by-step guide to making a sword. Well, now, sharpening is all that is left. Use a file or high-grit sander or sharpening stones and so on to sharpen it.

CHAPTER 3: BLADE-MAKING STEEL

Steel is made up of iron and carbon. All steels consist of other elements in limited amounts, including manganese, phosphorus, silicon, and sulfur. Steel is referred to as carbon steel if it doesn't contain any of these elements. Steel used for blades of knives are increased with extra elements and are referred to as alloy steels. These additions give unique properties for various kinds of steel. Alloy steels containing these additions are resistant to corrosion and labeled stainless steel, and when it comes to making knife blades, they're the most frequently used steels.

A well-built knife is a tool that functions consistently well without failing. Nevertheless, producing a knife that is not liable to fail can be a challenging task to accomplish, because you must sharpen a knife blade to a fine edge that must not dull or fracture. If you desire to achieve this, it is important to pick the right knife material as the wrong steel and grade will ultimately result in premature failure and edge dulling. Metals for knives are not all made in the same way. Below is a list of the best kind of steels for knife making.

Best Steel for Knives

Tool Steel

For knife production, tool steel is one of the most regular options. Generally, tool steels are carbon steels that incorporate additional alloying elements that boost their mechanical qualities. These alloying elements often raise the steel's resistance to corrosion too, however not to the level of stainless steel.

A standard tool steel grade that is utilized as a knife material is A2 (5 percent chrome, 1 percent carbon, air-hardening tool steel), which has excellent toughness, although can't range hardness to the level of some other tool steels.

Still, if not well taken care of, A2 could be susceptible to rust. Another solution that has higher resistance to corrosion and edge retention than A2 is D2 (high chromium, high-carbon, air-hardening tool steel). However, this solution gives a lower toughness. A top tool steel that is best at retaining a knife edge is M2 (molybdenum based), but when it comes to specific demands, it can be too fragile.

Carbon Steel

Steel grades with a high carbon quantity are attractive for knife making because they give the blade hardness and strength needed to hold up against impact and wear. However, the right heat treating must be achieved on high-carbon steels. If a quench is used too rapidly, the knife will be excessively fragile and may fracture. If the metal is normalized or strengthened, it'll be excessively soft, and the blade's edge won't be sharp for long.

Knives made from carbon steel can be inclined to rusting as well. This is due to the fact that carbon steel doesn't consist of many alloying elements that help to guard it against corrosion. It is necessary to ensure that a carbon steel blade does not rust.

Common grades of carbon steel for knife making consists C1090 (wear-resistance high-carbon steel), C1075 (high-carbon steel), and C1045 (medium-carbon steel).

Stainless Steel

This is the most frequently used type of knife-making steel, and it is the perfect steel for resistant-free knives. The additional advantage of using stainless steel is the chromium and other alloying elements inclusion that boost resistance to corrosion. Typically, stainless-steel knives are formed out of martensitic or austenitic stainless steels.

If you require a knife that has the appropriate edge retention, the ferritic and martensitic grades of stainless steel must have a sufficiently high-carbon grade, enough to achieve high hardness. Categories like 440 (high-carbon steel with the highest hardness and resistance to wear) and 420 (high carbon steel with at least 12 percent of chromium) have been commonly used in making knives.

The austenitic grades, such as 316 (standard molybdenum-bearing), is another stainless steel that may sometimes be utilized in making knives. Still, austenitic grades are typically incapable of hardening satisfactorily to give a lasting edge. Low carbon types of austenitic stainless steel, like 304L (extra-low carbon stainless steel), shouldn't be used when making knives, except resistance to corrosion is more important than the life of the blade.

Steel Properties

Choosing steel for specific knife making demands is based on the metal's properties and other aspects like manufacturability. If the metal is difficult to manufacture, then it is useless in a manufacturing environment. These properties are a result of the alloys included in the steel and by the techniques used in its manufacture.

Here are the ten most basic features of blade steel:

1. **Hardness**: A steel's standard and ability to resist sustained deformation. It is measured on a Rockwell scale (hardness scale dependent on indentation hardness of a material).

2. **Hardenability**: Steel's ability to be hardened by the process of heat treatment.

3. **Strength**: Steel's ability to endure applied power.

4. **Elasticity**: The ability of steel to flex or bend without breaking.

5. **Sharpness**: The blade's initial sharpness, functionality, and usability.

6. **Toughness**: The steel's strength to absorb force before the breaking

7. **Edge holding**: The steel blade's ability to keep an edge with no repeated resharpening.

8. **Wear resistance**: The steel's resistance to corrosion and wear all through use.

9. **Corrosion resistance**: The ability of the steel to resist degeneration that

may be caused by reaction with its surrounding.

10. **Productivity**: The simplicity and ease at which steel is machined, cold-formed, forged, extruded, blanked, and heat-treated.

Nomenclature of the Steel

The knife's classification, type of steel, and properties are often derived from the internal metal structure. Steel's internal structure suffer changes because it is heated and cooled. The systems based all through these changes are names such as Martensite and Austenite. Martensite is a stable structure that can be made by the rapid cooling of specific kinds of steel throughout the heat-treatment process. Metals capable of forming Martensite are commonly called martensitic steels, and this type of steel is common with the cutlery industry.

Additions for Alloy

By adding extra elements to the metal throughout the process of melting, the properties of steel can be changed. The alloying elements that are vital to knife production have been explained below these lines with short information on their effect on steel properties.

- **Carbon**: It appears in plain carbon steels, so it isn't an alloying element.

- **Chromium**: It's the main element in martensitic stainless steels commonly been used for cutlery utensils. It increases hardenability, resistance to corrosion, and resistance to wear.

- **Molybdenum**: It increases elastic strength, hardenability, and resistance to corrosion, particularly pitting.

- **Nickel**: It increases hardenability, toughness, and resistance to corrosion. It's one of the leading elements in austenitic stainless steel used occasionally in making dive knives.

- **Vanadium**: It increases hardenability and develops fine grains. The structure of grain in steel is a significant aspect of strength and wear resistance.

1095 Knife Steel

The primary form of carbon steel and the most commonly used in the creation of various types of knives is the 1095 steel. It contains 95 percent carbon, which reduces the measure of wear that a blade will encounter during its lifespan and enhances steel hardness.

Despite the reduction in wear caused by the high presence of carbon, the presence of the trace amount of manganese in 1095 steel makes it not as hardened as other steel types. Although manganese in higher concentrations causes steel hardening, it causes the blade to be more fragile.

1095 Steel Usage

The 1095 knife-making steel is extremely easy to sharpen and keeps an appropriate knife edge. But it tends to rust quickly because of its properties. These blade types will typically have some type of coating to resist rusting, but if the edge has been rightly preserved, rust should not be a major issue for you.

This steel is good for blades that aren't much thin because the steel is more brittle than other steel types. It's plain to sharpen; however, if a knife made with this type of steel doesn't have an appropriate measure of thickness in it, it can easily break. For this reason, it isn't the best grade of steel for sushi knives, folding knives, or tools.

It can be made to undergo heat treatment to increase its strength completely, but steel can end up fragile after this, breaking in the process. Although 1095 steel can be effectively utilized for chopping knives, it isn't the best option. Truly, it shines; however, there are more steels out there that are more appropriate for that application.

This type of steel is very suitable for polishing, although it hasn't been alloyed with stainless steel.

Characteristics

It can be used in producing replicate swords and daggers or blades. The military also makes use of it in their functional show swords. Not only is 1095 a major type of steel used in making dining tools, but it's also a more useful stainless appliance than the ones used in making swords. The 1095 steel is also a major steel used for rituals and in some religious ceremonies. Some kinds of machetes are also made from the 1095 steel.

Oils are essential in the maintenance of the 1095 knife. After every use, rinse and wipe clean, and oil it once every ten days. This gives the knife a polished look, and the oil also creates a barrier that hinders moisture from reaching the steel.

CHAPTER 4: COMPREHENSIVE GUIDE TO FORGING DAMASCUS

Damascus is an old method of craftsmanship that evolved from India in 300 BC. It started as a medieval culture in the western regions. Damascus steel has been proven to be exceptionally good and seemingly strange, capturing the hearts of many users. The art and craft of Damascus was a productive one and was made popular by the name of the region it emerged from. However, Damascus gained a flourishing industry in the area of weaponry as a result of the Arabians introducing Wootz steel into the city of Syria in Damascus.

Wootz steel was imported from Persia and Sri Lanka into Damascus for the production and utilization of hybrid steel blades. These hybrid steel blades have a characteristic feature of hardness and toughness. This technological evolution was clearly born out of intellectual minds who understood the resultant effect of merging various metals to create weapons of increased tensile strength over those made of pure steel. The unique trait of the Damascus steel is in its ability to captivate the mind yet remain mysterious although it is not pure.

The material has characteristics and features of multiple bands and molting welded together in a pattern-like model to make decorative blades of different shapes and length without easy annihilation. No record exists till today to prove the identification of alloys/intermixtures in original Damascus steel. Although in present-day Damascus steel making, the practice of merging pure metals is used with differences in relation to personal preference and needs.

Iron stands out in its use by metalworkers because of its carbide-enhanced solidity and grit, although the metalworkers have a wide range of choice to combine materials to make steel billets. According to the history of Damascus steel, early references were reported missing around AD 1700, leading to an inscrutable reputation for Damascus; however, this resulted in the downfall of patterned swords that later put a halt to production fifty years afterward. The booming industry that exists today has constantly put to remembrance the significance of the ancient tradition of the Damascus steel.

Steel Composition

Metallurgy and chemistry are the fundamental basis for the alloy/composition and the various applications in steel production and utilization. The ingenuity and technicality of Damascus techniques and processes are directly linked to the sciences behind Damascus steel and steel making. Due to the durability and longevity attributed to the former reference, the Damascus steel was referred to as the superplastic, not because it was not a real metal and it was never pure. Various types of modern steel have exceeded the Damascus blades in terms of efficiency; moreover, the blades produced recently have been extremely rigid and effective due to its native chemical composition. Different steel types welded together to form billets have been used in the production of modern Damascus steel blades since 1973. These billets constitute strips of iron that aid firmness on a molecular basis. This enables flexibility in layering them out according to the needs indicated by specific utilization of the blade and preferences of the blade owner. Damascus steel blades are custom-made to suit individual preferences; they are not produced in a rigid like pattern.

It involves a simple process:

Steel ingots are formed into billets. The billets make a sandwich-like fold pattern in other metal types. The product formed is made up of a series of layers in hundreds or more with good varied design and solid density. The steel has a high level of integrity and uniqueness as a result of this tested process.

Ductility and brittleness are two major dichotomous structural types that make up the fundamental composition of the Damascus steel. Ductility aids compression of the material in order to absorb high levels of energy that helps to reduce failure in the integrity and efficiency of the blade. Brittleness, however, relates to feebleness, which is highly misleading.

Brittleness refers to the flexibility of the material in the prevention of breakage, also in enhancing edge sharpness. The Damascus blade has an advantage of durability and easy cutting due to this structural process. The convex grind provides sharpness to the thinness of the edge so that sliced material yields to the sides during the stroke and thus reduces "sticking" that often occurs with blades having blunter edges. The convex grind is aided by the structural brittleness.

During the process of forging, malleability and sustained strength are acquired due to the formation of carbon nanotubes in the steel. High performance, efficiency, and strong quality in steel integrity are ensured by a high concentration of carbon. This is the sole reason why carbon is essential in the development of Damascus steel blades. During the process of forging, small steel ingots slowly form into a preferred shape of a blade. This causes the iron carbides to align into bands that form distinct patterns. These patterns evoke feelings in the Wootz steel from ancient India, and they portray old aesthetics and style of production. Metalworkers today are able to replicate much in the same order in which Damascus steel was known to exist centuries ago.

Heating and Finishing

The distinct details of the Damascus steel vary according to individual preferences and choices and also on the type of metals being merged together. Below are the lists to the general heating and finishing processes for preparing Damascus steel. A fixed temperature between 1,500 °F and 2,000 °F is used in the heating treatment of the steel in relation to the banding and a combination of austenite and cementite.

- Fix the temperature of the furnace according to specifications.

- Place the metal block in the furnace. Heat through its cycle to its initial temperature.

- After heating, cool steel by soaking for ten minutes.

- After quenching steel in oil, transfer to liquid nitrogen for 60 minutes.

- Steel tempering involves subjecting the steel to heat and cold in order to make it strong and hard. This process must be carried out twice.

Steps in the subsequent finishing treatment:

- Apply a grit finish to the blade.
- Etch in 50/50 diluted solution of ferric chloride and distilled water without pre-buffing. Leave the blade in solution for at least ten minutes.
- Remove the blade and rinse under running water.
- Repeat the process at five-minute intervals till you get your preferred result.
- To neutralize finish, immerse blade in tri-sodium phosphate.

Making of the Damascus Blade

Prior knowledge of heating and finishing, the Damascus steel is highly needed in the making of the blade as shown in the process discussed below. Therefore, a metal maker needs a detailed knowledge of both processes in order for him to complete the process below. It's an essential order for specifications, and the process works together.

The process of making the Damascus is simple and easy; however, it must be done painstakingly and with caution. Also, it's a time-consuming process that ensures real aesthetics and the development of a well-balanced, productive blade.

1. Glass and leaves materials are added together in a crucible. These help to prevent oxidation.

2. Heat crucible to melt the materials together.

3. At the cooling temperature of the crucible, carefully remove the metal ingots and heat them to a specific temperature required of forging. This stage entails "sandwiching" of the process described above. The metal is hammered while it is still very hot. After metal completely cools, reheat the metal to forge again. This cycle is repeated to sharpen edges and shape the blade.

4. Cut the blade and hand-forge the final details after the final shape is achieved.

5. Shave away the excess carburized metal from the surface of the blade.

6. Insert grooves and drill holes into the blade surface as needed or desired.

7. Hammer blade flat again by reheating. Polish to set the blade's near-final form.

8. Etch the surface of the blade with acid to accentuate the pattern.

9. After completion, clean acid thoroughly from the surface of the blade.

Damascus Knives

There are various types of Damascus steel knives, and they are used for a wide range of purposes—for example, in wood cutting, hunting, camping, and so on. The reason for use and the type of knife determine the composites that are merged together to make a Damascus knife.

Examples of some common types are listed below:

- Tactile folding knives
- Flip-flop knives
- Hunting knives
- Carving knives
- Serrated knives

- Rigging knives
- Tactile fixed blades

A unique and distinct feature of any type of Damascus knife is in its endurance—that is its ability to endure force of any kind applied to it. In addition to the enduring force, it also has a long lifespan.

The aesthetics and high efficiency of the Damascus steel knives are its most common aspects; however, lovers of the knives see it as the perfect example. There is an absolute distinction between two Damascus knives as no two knives ever look the same. Damascus knives have fashionable patterns engraved by metalworkers into the blade during the forging process. Each knife is unique and priceless as an expression of art. In view of its mysteriousness and special quality, the Damascus style will undoubtedly last longer in the years to come. Even after two thousand years, the tradition and style are waxing stronger.

CHAPTER 5: LOST WAX CASTING

In lost wax casting, the wax pattern is melted in the process of molding, and it's an old-fashioned technique. Hence, the name lost wax was given to the process. The attracting force of the lost wax to various foundrymen and artists lies in the integrity and dependableness of the method. This process also enhances the formation of elaborate sculptures, ornaments, and automobiles. Despite the fact that the name of the process has been changed to investment casting and several other names, the method still maintains its luster and glow.

Lost Wax Casting Process

In the ancient times, hands were used in making the lost wax casting for every single piece of mold. However, in the modern world, one pattern can be made into a series of patterns, although the pattern is melted or lost during the process. Hollow patterns or solid can be used for molding purposes

There are two major ways of making the pattern: direct method, which involves directly making the wax, and indirect method, which is by sculpting the model in claylike materials and then making a wax copy out of it. The molten wax is then poured into the mold to the required thickness or consistency and turned upside down to cool. In order to give a flawless finishing, the wax pattern is carefully removed from the mold and improved upon after cooling the wax. Spurs and gates of wax are joined together for pouring the molten metal after the wax pattern is ready. When the pattern is wet, materials enclosing the pattern will hold it firmly in shape. A grainier material is added to make a solid mold in which the metal can be poured, and it then solidifies.

The mold in the kiln is heated till the wax liquefies. Mold placed in the kiln is fixed in place with flasks. A vacuum is then formed in the cavity where the wax existed. The molten metal, like bronze or gold, takes its place. The mold is detached after the metal cools in order to extract the cast.

Applications of Lost Wax Casting

- It is used in the casting of necklaces, earrings, and other small parts. Brooches and buckles are clothing accessories that are also made from the lost wax casting process.

- In the making of engine blocks by automobile producers, a method known as the lost foam is often applied. In recent times, the lost wax process has been often used in several other areas, like dental restoration, fine jewelry, and sculptures. Silver, gold, aluminum, brass, or bronze are cast with this method.

Merits of Lost Wax Casting

Flexibility of material:

- Any material liable to burning, evaporation, or melting can be used with the lost wax process to create a mold cavity.

- It also replicates the fine details of the initial wax.

- It provides shapes of casting that would be difficult in other methods.

CHAPTER 6: JAPANESE BLADE

The art and craft of forging a Japanese sword is a meticulous and cautious process, which has improved over the years both in stylistic and aesthetic considerations, as well as improvements in technicality. For the absolute beauty and flawless finish of these blades, the smith should have certain qualities, which include finesse, patience, and a detailed eye for the beauty of the materials, as well as the limitations.

Tamahagane was often used by the Japanese smiths in ancient times. The tamahagane was a kind of steel made in a Tatara shelter from iron-rich sand. Although modern-day smiths still use this type of steel, the ancient manner of making swords by the Japanese is now made in the last functional Tatara shelter located in Yokota, Shimane Prefecture.

Although the Tatara smelting process is highly efficient, it still has some disadvantages which include the following:

1. High concentration of impurities
2. Lack of consistent dispersal of carbon content, which is a vital tool for converting iron into steel

When there is a high concentration of carbon, the metal will be fragile, and when there is a lower concentration of carbon, the metal will be too soft.

Kitae: Forging the Blade

The folding technique of kitae was majorly developed in the bid to correct and balance up for the quality of the tamahagane. The smith will firstly choose good pieces of tamahagane and forge weld them into a single block. This single layer of block forms the outer part of the finished blade. After that, the smith begins to hammer out and fold the block back on itself. This is a very strenuous process. There are two main resultant effects of this process:

1. There is a standardization of the carbon throughout the metal and the extraction of impurities out of the steel. With a great wealth of experience, a smith can direct with precision the quality of the steel in this way.

2. The folding results in the patterns that have made these blades popular all over the world. Layers are formed every time the block is hammered and folded backward. A 14-time fold produces 16,000 layers. The surface

in between the edge and ridgeline now concisely shows texture in the *ji* when the blade is finished. By contrasting the direction of the folding, the smith has a choice to choose any specific texture he wants, such as *ayasugi-hada* (concentrically curved grain) (2001.574) or a *masame* (a straight grain parallel to the edge). He can also fold the block continually in the same direction in alternate directions or crosswise. Each method results in a unique fashion of texture.

The *kawagane* is the outer skin that envelopes a softer core, or *shingane*. This combination enhances the flexibility and the strength of the blade to withstand breakage under stress. The harder *kawagane* also suits sharpening than the more ductile core. The two layers are heated and hammered out into a long bar. This welds the layers together and forms the blank from which the finished sword is made. After the blade has been forged into its fundamental form. The smith uses files and planes to bring out a final shape after a rough polish.

At this junction, we can accurately pinpoint the distinguishable features of the sword, which includes a clearly defined profile, point, and ridgelines, as well as the tang and a smooth level surface.

Yaki-Ire: Hardening the Edge

The most complex and significant aspect of the sword-making process is the hardening of the edge. The super quality sharpness of the blade is as a result of effective hardening of the edge.

The blade is firstly coated in a mixture of water, clay, ash, and other ingredients. This mixture is known as *yakibatsuchi*. Every smith has his own unique recipe, which is often hidden from others. The smith then spreads the yakibatsuchi systematically over the surface, creating a thicker texture along the spine and a thinner texture at the edge. The smith meticulously heats the blade while working in a forge room without light, except the light emanating from the glowing of his coals.

Crystal structures in the metal undergo changes as the temperature increases. The smith carefully observes the color of the glowing blade, and when the high temperature is reached, the sword is quickly dipped in a trough of water.

Steel structure changes to austenite at the critical temperature, around 750 °C, a phase where carbon and iron combine completely. When the blade is rapidly cooled by quenching, austenite becomes martensite, which is the hardest kind of steel.

Where the thick yakibatsuchi was applied, however, the blade will slowly cool down, transforming into ferrite and pearlite instead of martensite. These are softer and more flexible. The combination of softer body and hard edge is what gives the blade its desirable qualities, just like the *kawagane* and *shingane*.

Also, the hardening of the edge makes an obvious change in the metal's surface. A variety of effects can be produced depending on the manner in which the clay mixture was applied. This edge design is known as the *hamon*, and it is one of the most vital aspects in the blade's aesthetic appearance. Every one of these patterns has a specific name. For instance, *sanbon-sugi* is a zigzag line in clusters of three while *suguha* portrays a very straight hamon.

Following the edge's hardening, if the smith is happy with the blade's quality and appearance, it is then passed to the polisher, who then gives the blade a last mirror-like polish, and other craftsmen, who will then make the sword mountings and scabbard.

Complete mountings (36.120.417,418) have numerous elements, including metalwork like *tsuba* (36.120.79) and *menuki*, wrapping, lacquered wood, silk cords, and ray-skin grips. Though all these are pieces of art in themselves, the blade is the only true centerpiece of the completed work, an instance of the ingenuity that has been shown by Japanese smiths for centuries and their desire to accomplish the ideal blend of art and technology.

CHAPTER 7: GUIDE TO BUILDING A SIMPLE FORGE

There's one thing you have to face if you desire to work with metals—heat is needed. With the availability of heat, you can make any metal, no matter how tough, submit to your will. You'll never gain complete mastery over this stubborn material without heat. However, it isn't difficult to finally make the decision to take another step toward teaching yourself this smithing skill.

You'd learn how to build an effective but simple forge in this section. With this forge, you can heat steel hot enough to the temperature needed to effortlessly shape it.

Step 1: Materials

Some rocks, like granite, a bucket of mud, and an iron or steel pipe, are required for building a forge. Due to the fact that heating galvanized metal is bad, do not give consideration to galvanized pipes. However, coating the pipe in some layers of mud and using the forge outside for proper ventilation can be used to counter it.

Step 2: Construction

Place the rocks in a ring. You can create it any size you desire, but it is preferable to keep it small so as to get maximum heat.

After this, you need to use a thick layer of mud to coat the inside of the forge. Coating with mud will prevent the ground from absorbing the heat. The rocks are also insulated by the mud so they don't crack.

Step 3: First Firing

To get the fire sufficiently hot, the metal pipe should be connected to a vacuum cleaner on blow. Some steel can be heated to test how hot it gets and then melted in the can.

Step 4: It Gets Hot!

The forge at this point reaches forging temperatures able to melt steel.

Step 5: Conclusion

The forge can be upgraded a bit and made bigger.

CHAPTER 8: ARRANGING THE SHOP

Make your tools and shop neat with a place created for each thing. Take proper care of your tools and plan your shop arrangement.

Arrange your tools so that they are all in a tight little row with a high, adequately spaced workbench just opposite them. This enables easy turning around from any tool to practically be at the workbench at any time.

Each of the small tools should go in a set of drawers directly under the workbench.

Next to the belt sander, place a low, strong table to pile the belts on and do odd jobs. It could also be useful for riveting and welding. Ensure that it has a piece of railroad track and a manual punch press bolted to it. To place slots in metal pieces for hilts and finger guards, the punch press should be used.

Then you could have a joiner (wood planer) and circular saw at the side out of the way of your regular knife-making area. These tools are used to mill down bigger pieces of wood for making the knife handles.

Having every one of these tools arranged and set up in this way comes in extremely handy for doing numerous other things other than knife making.

Lighting

You will discover that fluorescent lights provide better area illumination and are much more efficient than incandescent bulbs. The incandescent should be used to spotlight the working region of each of the machine. Fluorescent light fixtures are less difficult to locate in second-hand shops that specialize in interior fixtures.

LEAVE A REVIEW?

Throughout the process of writing this book, I have tried to put down as much value and knowledge for the reader as possible. Some things I knew and practice, some others I spent time to research. I hope you found this book to be of benefit to you!

If you liked the book, would you consider **leaving a review** for it on Amazon? It would really help my book, and I would be grateful to you for letting other people know that you like it.

Yours Sincerely,

Wes Sander

CONCLUSION: SELF-EDUCATION

Free lunch doesn't exist. No one is going to teach you everything about a craft just because you ask them to. You may be *extremely* lucky to find somebody who is willing to share their hard-earned knowledge, but they will have expectations that *you* have done your part.

Very few books were available on the craft of blacksmithing in the first half of the twentieth century, but none was available on bladesmithing. About the time the blacksmithing started dying, people in their large numbers started having a renewed interest in it. The outcome is that there are now numerous excellent books on the subject of bladesmithing and knife making.

If you will do any form of bladesmithing, patience and ingenuity are as crucial as inspiration and resolve. Try not to burn yourself out by attempting the actual knife making before you are ready. Prepare yourself as much as you can and take your time about it. Your knives' quality will reflect the way you went about making them. Your evenings should be spent paging through metallurgy and tool-making books.

In the long run, you would realize that bladesmithing is beautiful and the adoration of its maker is contained within its vibratory structure. You would understand that you are putting this affection into the food you eat and the work you do. This consistent subtle reminder of love combined with your everyday food and work can infuse a heightened level of spiritual warmth into a life that might have otherwise been of a mechanical and impersonal nature.

Bonus: Now that you have made a knife that you are proud of, wouldn't it be amazing if you could sell it too?

Get my knife selling secrets with my **100% FREE** e-book, *'Bladesmith's Guide to Selling Knives'*.

All you have to do is go to **http://eepurl.com/gbq6Sb,** and enter your email and name.

Thank you and good luck!

Made in the USA
San Bernardino, CA
11 March 2019